Let's Get Healthy

Your Body

Let's Get Healthy

Your Body

W

FRANKLIN WATTS
LONDON • SYDNEY

This edition first published in 2008 by Franklin Watts.

Franklin Watts
338 Euston Road
London
NW1 3BH

Franklin Watts Australia
Level 17/207 Kent Street
Sydney NSW 2000

Let's Get Healthy is a reduced text version of *Look After Yourself!* The original texts were by Claire Llewellyn.

ISBN: 978 0 7496 8322 1
Dewey Classification: 613

A CIP record for this book is available from the British Library.

Printed in China

Series editor: Sarah Peutrill
Design: Kirstie Billingham
Illustrations: James Evans
Photographs: Ray Moller unless otherwise acknowledged
Picture research: Diana Morris
Series consultant: Lynn Huggins-Cooper

Acknowledgments:
Biophoto Associates/SPL: 23tr
Layne Kennedy/Corbisstockmarket: 19
Brian Mitchell/Photofusion: 23bl

With thanks to our models: Aaron, Charlotte, Connor, Jake, Holly and Nadine.

Franklin Watts is a division of Hachette Children's Books, an Hachette Livre UK company.
www.hachettelivre.co.uk

Contents

Looking after yourself

Your body is working all the time. You need to look after it.

Treat your body well.

When we are very young, our parents look after us.

As we grow older, we can do more for ourselves.

We learn to care for our bodies.

Keeping clean

You need to wash
when you get up in
the morning.

Remember
to keep your
hands and
nails clean.

By the evening, your skin is dirty again.

Take a shower or a bath.

Wash away germs

Germs are everywhere - on our bodies and on the cat!

Where?

Germs are tiny living things we cannot see.

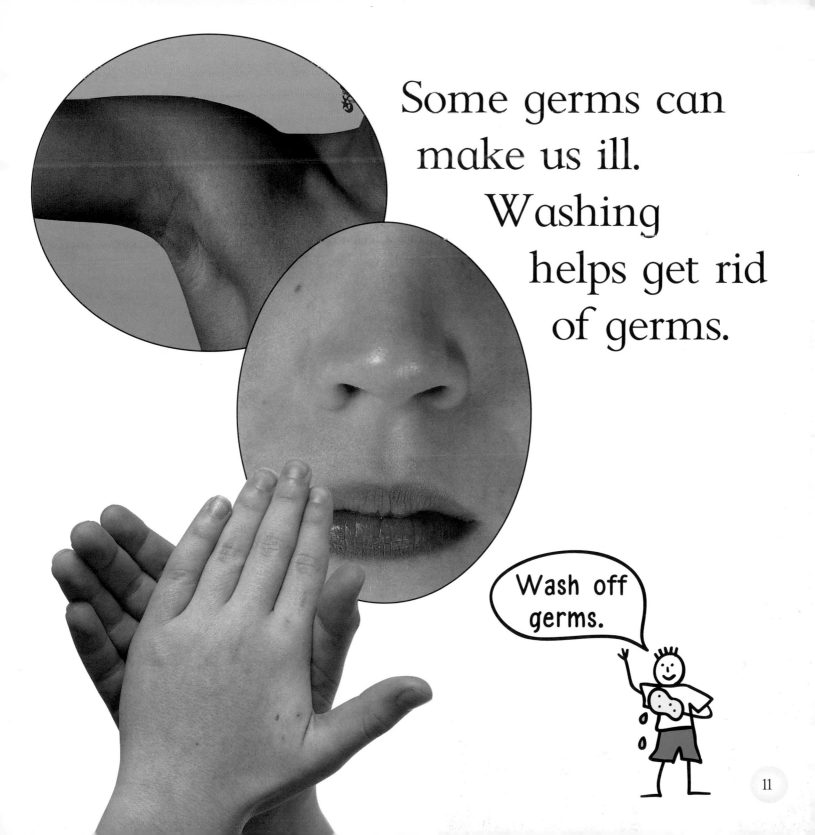

Some germs can make us ill. Washing helps get rid of germs.

Wash off germs.

11

Teeth, hair and nails

Brushing your teeth helps keep them clean. It also helps stop tooth decay.

You need to
wash your hair
once or twice
a week.

Short nails
are easier to
keep clean.

Germs can spread

Germs can spread from person to person and make you ill.

Washing your hands with soap can stop germs from spreading.

Use a hanky to stop cold germs from spreading.

Always sneeze into a hanky.

A healthy diet

A good diet is important.

This group of foods gives you energy.

This group helps to keep you well.

These help your body grow and repair itself.

I eat foods from each group.

Sugary foods harm your teeth.

Try to eat less sugary foods and choose other snacks instead.

Keeping fit

Exercise is very good for your body.

Skipping gets you fit!

It makes your muscles bigger and your bones stronger. It helps make you fit and strong!

I'm strong!

A good rest

Our bodies like to exercise, but they also need to rest.

Sleep rests every part of your body.

It makes you feel good as new.

On the mend

The body is good at mending itself.

Ouch!

You can help it by keeping cuts and grazes clean.

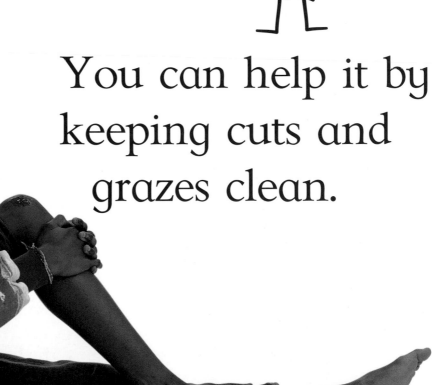

22

This x-ray shows broken bones.

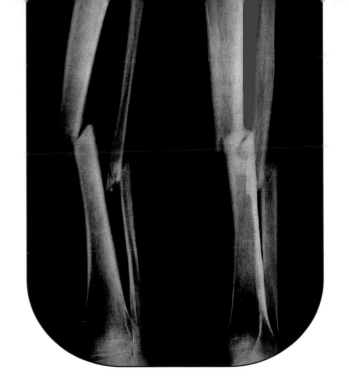

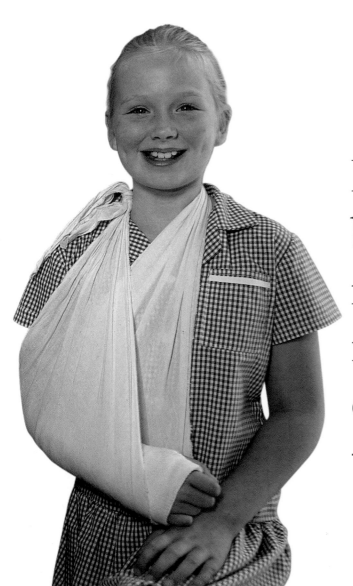

If you break a bone, a doctor sets it straight. You may need to wear a plaster cast while the bone mends.

Feeling ill

Rest and sleep help you get better from illness.

I'm resting in bed.

Sometimes we need
to see a doctor.

They can help us get better.

Keeping safe

Sometimes we need medicines to make us well.

Never take medicine unless a doctor or an adult gives it to you.

Grown-ups keep medicines locked away, and in high cupboards!

Glossary

diet The food and drink that you usually eat.

doctor A person who tries to make ill people better and keep well people healthy.

energy The power we get from food, which makes us able to work and grow, and keep warm.

exercise To be active.

germs Tiny living things that can spread disease and make you feel ill. Germs are too small to see.

graze When you scrape your skin.

healthy   Fit and well.

illness   Being unwell.

medicine   A liquid or pill that you take to make you well.

plaster cast   A mould that holds a broken bone in place, so it can mend.

sugar   Something that is found in many foods and makes them taste sweet.

tooth decay   When teeth rot and develop holes.

x-ray   A special photograph that shows the bones inside you.

Index

About this book

Learning the principles of how to keep healthy and clean is one of life's most important skills. **Let's Get Healthy** is a series aimed at young children who are just beginning to develop these skills. **Your Body** looks at cleanliness and keeping the body healthy.

Here are a number of activities that children could try:

Pages 6-7 Discuss all the things they can do on their own. How much has changed since they were babies?

Pages 8-9 Test some different soaps and shower gels. Which creates a lather most quickly? Which has the nicest smell?

Pages 10-11 Make a poster to explain the ways we can avoid germs.

Pages 12-13 If appropriate, look at and test different nail scissors and cutters. Which is the easiest to use?

Pages 14-15 Write about the last time they were poorly. How did they feel? What and who helped them to get better?

Pages 16-17 Write a menu for one day. Make sure it contains a variety of foods from all the food groups.

Pages 18-19 Devise and play some games with friends that use lots of energy. Afterwards, decide which game was the most tiring. Why?

Pages 20-21 Find out how many hours of sleep different people in a family get. Who has the most sleep? Who has the least? Discuss why this is.

Pages 22-23 Discuss why cuts, grazes and other accidents can be painful - the body is warning that something is wrong.

Pages 24-25 Make a list of all the qualities a doctor needs. Which is the most important?

Pages 26-27 Discuss where we get medicines from (i.e the chemist).